Notes to Self

thoughts along the way…

W.J. Hamilton

Armchair Books
3317 E. Bell Rd. Phoenix AZ 85032
www.armchairbooks.net

ISBN: (sc) 978-0-9976522-2-2

ISBN: (e) 978-0-9976522-3-9

Library of Congress Control Number: 2016948849

Armchair Books Services rev. date: 08/5/2016

DEDICATION

This book is dedicated to my granddaughter, Charlotte Amelia Carter. It has been years in the making but, Charlotte, your arrival has given me the courage to 'bind' some of my thoughts and images into this form. Every word and image on the following pages are a gift from God.

I am grateful that *you* have pushed me into these uncharted waters. Thank you, I needed a new adventure.

Love is a powerful motivator and I love you very much.

(*For the record, I have not edited or altered any of the original text to make it more palatable or eloquent. My original intent was to be honest before the Lord-not to impress or persuade.*)

Granddad

July, 2014

CONTENTS

Christopher Creek, Arizona

Anyone who has spent time in the wilds has had the experience of walking through the woods, thinking those woods are empty. Then they stop to rest, sit silent for a time, and the woods slowly cease to be empty: animals, birds and insects, botanical and geological nuances, movements of cloud, sound of wind, unseen water, make themselves known. A home place is like this. "Homing" is an interaction that begins when all other action ends. We must be going nowhere in order to be somewhere. Sense of place is received, not aimed at. – David Duncan [1]

INTRODUCTION

In this short profound observation, David Duncan has described what has been happening to me for 40 years. *Homing* for me is an interaction, a communion, with God that begins in my spirit when *all other action ends*. Over the years when I have allowed myself to take the time to be *nowhere* words have come through my pen from *somewhere* that have proven to be a blessing to my soul and to some friends as well.

For what it's worth, I have compiled a handful of words that were *not aimed at* but have been born in a space, a *home place*, where God faithfully chooses to meet me.

Might these words that I am offering here encourage you to go *nowhere* so that you might hear God's voice *somewhere*.

1

WORDS

November 21, 1995

I am more ‘honest’ as a writer [in written form] than as a thinker [cognitive form]. I cannot write as fast as I think. Consequently, the pace of writing forces my mind to be more present and in touch with my emotions and feelings. Writing enables me to *hear* the Lord’s voice.

November 12, 2004

I realize that some of the joy, connection, peace etc. that I enjoy from writing in my journal has also been experienced by friends who have received letters from me. I can't get over the fact that when I take the time to write down words/thoughts they have often made a powerful and or profound impact. Something happens through my pen that doesn't come spontaneously or so naturally through my mouth–my words. And of late, I have written a handful of letters that have communicated some powerful truths that would not have been as effective had I spoken them. In fact they would have never *been*. Only when I stop and begin to write do thoughts come through my pen that were never formed as words on my tongue. It seems that the letters on the page were not meant to be *heard*. As if they would not, could not, exist as spoken words. This is a mystery to me. It seems that writing is my best form of expression – most powerful tool for me to expose my heart. This expression is not necessarily just for other’s edification but for my own as well. I learn about Jesus and his power through these thoughts. Often it is new and revealing stuff and I am encouraged by the revelation. It seems odd that it was not known to me until it came out as text. I am often blessed and amazed by some of the concepts and phrases that I have penned. Not amazed in a proud sense like, "Wow, what a deep thought, I'm so smart or clever.” But, more it is a sense of humility like, "Wow, where'd that come from? I did not know I *knew* that." I can't explain it but I am so thankful. In writing I seem to commune with Jesus in a wonderful and personal way. And I sense he is glad to communicate to me in this way. What a privilege.

South Rim, Grand Canyon, Arizona

March 23, 1998

Yesterday at church a man was using sign language to translate for some deaf people. I couldn't stop watching him. I was moved by his *words*. I don't know why it was such a mesmerizing experience. Was it his face? He had such a look of awe, peace, power, compassion all in one. Maybe it was the *words*. His emotions during the songs compelled me to want to weep. Would a normal conversation or a secular song in sign language do the same thing to me? I had never *heard* so well before. *Seeing* words while hearing them is awesome. It fills in a *gap* that I never knew existed. Hands make words bigger and more meaningful, like silence and solitude make a stream sound louder.

January 29, 2008

I'm having some thoughts about the future and the use of the *written word.* Words on paper seem to be losing steam, becoming irrelevant. Text from a handheld or phone or computer seems to be the way our culture will *write* in the future. It would seem books and letters may become obsolete and disappear. I love to write and I prefer to read words from pages rather than from a 'screen'. I have been sort of fearing this loss but I don't think written words will disappear. In fact, I believe they will always be with us. Machines and wireless gadgets 'connected' to somewhere can crash and be lost, their words can disappear, evaporate. Oh, the sense of loss and powerlessness we experience when there's been a *crash* with no way to retrieve the meaning, the message. But books and notes and journals will prevail if they avoid fire and mildewed boxes. Words that are connected to paper have a better chance to endure than megabytes of letters lost in *space*. Bring on paper and pens.

2

WATER

April 14, 1996

The sound of birds is around me, the sight of them as well. A red breasted *something* is catching unseen insects above a pool. An occasional trout breaks the surface and fills its breakfast menu. But, the most wonderful part is the sound of the Lord's voice in the rushing water of a tributary flowing in. The soothing calm *voice* that says, *"Come and sit, relax. You have come away from your busy life, you have made the effort and I am your reward. Sit and listen. Don't feel guilty – sit and listen. Don't think of home, of troubles. Feel the Father's presence. His voice is constantly coming out of the trees and falling over the rocks. It was here before you came and it will continue to speak long after you leave. Listen... Don't wait so long before you come again. And remember, you can sit anywhere and listen and I will be there."* Yet, I think His voice is clearer at the water's edge.

West Fork of Oak Creek, Arizona

January 11, 1993

I'm sitting here in my car at Sabino Canyon. It's beginning to rain again. The desert is officially saturated from many days of rain, all the ravines are flowing. This water has nowhere to go but downstream. The rivers will surely rise.

Oh, to be so *saturated* with the Spirit, with Jesus, that each time He *rained* down on me I would overflow and drastically affect what's *downstream* instead of just soak him up and keep him for myself. Lord fill me!

September 2, 1998

The water sounds soothing. It's muddy, a clear sign that there is weather upstream. But, that hasn't affected the sound. It would be nice to believe that regardless of my *condition*; muddy, clear, etc. my 'sound' would remain the same. I think that would honor God.

Grand Falls, Arizona

April 5, 1999 Greer, Arizona

It's 6:45 AM and the world is very still outside my window heavy laden with snow and cold from the long night. Yet, the stream, the Little Colorado, is moving past. I guess it's not too cold. It's doing what streams are made to do; pass by my feet and eyes to delight me and then never return. A short-lived moment of peace for me. Is this enough to validate the stream's existence? I think yes. Its job is to serve me the watcher. But, not only me, the countless other 'watchers', peace seekers downstream. This stream will continue to flow constantly bringing peace in its color, sound, shape; it's *being.* Will it tire of this holy work? It cannot. It only knows how to flow one way. Does it know the joy it brings with its passing? No one can say. But, if it does know, I will thank it and every stream I find myself being blessed by. I will thank the Father who created the water that fills each stream that I have and will stumble upon. He gives great gifts; like this quiet stream passing through this cold surreal painting. I will thank the Creator and he will be glad and the water will continue to serve him and all who venture to its edge downstream from my communion moment.

Ribbon Falls, Grand Canyon, Arizona

October 22, 1999

If you stare at a waterfall it looks like a constant flow, one big, connected, powerful fall. But if you pick out a drop of water at the top of the fall and follow it all the way until it splashes in the pool below, it looks as though it is in a suspended state of animation. Like it is moving in slow motion – almost like it stops during its fall and looks around. It seems to be aware of itself as it's falling to the pool.

Could this be a word for us: to live in the now; to be present and aware of our self during the rush of our life? It is a purposeful fall, not optional for the water but yet, it can still enjoy the fall; the ride. We choose our falls (though some would argue to the contrary) and when we are falling, we can choose to be aware as well. What a view, what a ride, what a splash! Yes, we need to pay attention during our falls. What a cool opportunity.

June 5, 2011 Laguna Beach

The tide is coming in. When the tide goes out so much gets exposed – a wonderland of hidden rocks and reefs and the creatures that inhabit them. It's a short opening that allows people to get a glimpse, it seems to draw people out to look and explore. Explore for what? For what is not normally seen, the *hidden*. We seem to yearn for what we can't see and then get drawn *out* where we get exposed looking for *something*. Then, before we know it, we are chased away by the natural flow of nature and resume our higher tide *position* of sitting or biding our time until we get drawn out to seek and explore and get exposed again. I think we should get in the habit of seeking and exploring during the *high tide* hours as well. What might I be missing that is up on the beach (exposed all the time) that the Lord may want me to see? Are there *creatures* surrounding me that need discovery? Am I only looking occasionally when I should be in a 'chronic state of wonder'?

Maui, Hawaii

3

Evangelism

February 16, 2004 Being right can be so wrong—

Two nights ago Jan I sat with some other couples discussing the book we had all read. During the discussion one couple made it very clear they felt the author was, theologically, way off base. To a person, no one agreed with them but everyone was quite sensitive in trying to help them see how they were overreacting and saying something that was clearly not in the text. The thing that really got to me was the self-righteous, holier than thou, I'm smarter than you, attitude from the couple, especially the husband. Here is a man who teaches Bible systems to people in Latin America through an international evangelistic ministry. He's the expert. His attitude and demeanor were repulsive to me. It really got to me, deep in my heart. Janny said I did a good job of communicating and not condemning. But in my heart I was angry.

I've been trying to sort out what was motivating the anger. For one, I feel sad for the men and women he is teaching who are being led to believe in systems that often can lead to condemnation and judgment as opposed to Christ-likeness. It's a righteousness/works concept that can steer these faithful students away from embracing the mystery of God's amazing grace and love. This sadness doesn't overwhelm me because I trust that the Holy Spirit can do what he wants in the hearts of those who call upon his name, regardless of the way they are introduced to him. But, still the thing that really burdens me is the husband's attitude. He is so self-righteous and demeaning. He had this thin, little, prideful, condescending (I'm the Bible expert) smile on his face during the entire discussion. I'm sure his goal was to communicate the truth, but who would listen? As he defended his position Saturday he seemed to not care that the people who were in the room were well educated elders of his that had long term relationships with Jesus: a college professor, a seminary graduate, and several who had walked with Jesus for 40+ years. How could someone show that much disrespect? I believe he would say, for the sake of the truth, and its defense, it is okay to disrespect a person as you attempt to convince him/her to believe. If the ultimate goal is to convince/convert then respect can get trumped by this theological zeal.

But, that is not how Jesus did it. He did not convince with words or strong theological argument as much as he did with his *life*. He walked with the lost and ate with sinners and used friendship to teach about *the truth.* He was more concerned with *being* an expression of the Father than talking people into following him. I never want to be 'that guy' on the other side of the room thinking I'm right and secretly shaking my head at the poor lost bastard across from me. Lord, help me love those who I think are *wrong*. And please shine your light on my *wrong* so that I never encourage anyone to believe in any*thing* other than Jesus.

February 23, 2006 *My priestly confession...*

Father Maguire died a couple of days ago. He was 92 and served at Our Lady of Perpetual Help Catholic Church for over 50 years. When I think of him I picture a "typical priest": bound by duty and call, committed to his job and functioning mostly out of obligation. I have this twisted view of priests not really liking their jobs and being mostly sullen but committed to serve and administer the Holy sacraments. Basically, I can discard them as men bound by religion and therefore consequently ineffective and powerless on a personal level with people. Their *clan* certainly doesn't have a strategy of evangelism.

Now that this *confession* is out of the way, I can share a new revelation the Lord is germinating in my heart:

Is it possible that, during my childhood, when Father Maguire would occasionally come by our house and sip whiskey with my dad that it was some kind of evangelistic strategy? Could he have been '*earning the right to be heard*'? For the first time in my life, just two days ago, I had the thought (I believe it was from the Lord) that this Irish priest was sitting with my dad out of a desire for him to come closer to God, that it wasn't out of some holy obligation or just to get a buzz. Father Maguire was *caring* for my dad – 'witnessing' as it were.

And, could it be that my dad was, purposefully, caring for the priest by offering him a sanctuary away from the lonely world he lived in, surrounded by nuns and duty and religion?

Were they both caring for each other, doing their best to muster up concern and love out of their own poverty and bless a fellow brother as they raised their glasses together?

Were those moments they shared on Cheney Road *holy* moments?

This is my postmortem apology – confession to Father Maguire: *Bless me Father for I have sinned... Forgive me old priest for my judgmental spirit and lack of wisdom and discernment. I am sorry that I, all too often, limit the power and ability of the great God we both love and serve. Thank you for loving my father and sharing those holy moments with him. Very few took that chance with him. Rest in peace.*

Mogollon Rim, Arizona

July 24, 2011 *Name change…*

When I was 12 years old, my mom remarried and my step-dad adopted us kids. My last name was changed from Peil to Hamilton. On the day the change was official I decided to stop being a juvenile delinquent.

I have used that *adoption* story and subsequent changed name in talks as an example of the transformation that takes place when one adopts, or is adopted into, the Kingdom of God. This morning I am realizing that my motive for behavioral change after becoming *Hamilton* was based mostly on fear. Bill Peil was afraid of being arrested. How often have people come to Jesus motivated by fear of their *name,* their past, as opposed to being won over by His great love for them? Is it too much to think that *evangelizing* by sharing the love of God without any fear tactics or threats of eternal punishment is a healthier path to God?

South Rim, Grand Canyon, Arizona

4

WHERE HOLY SCRIPTURE HAS COME TO LIFE

I learn best from stories.

A good story draws me in and at some level offers me space on the stage. Sometimes I take the place of an established character and other times I expand the cast and write myself into the scene. There are a number of stories in the Bible that are so compelling and fascinating I find myself going back to reengage with them on a regular basis. At some level, it seems unsophisticated or juvenile to reread a biblical story over and over again. A critical voice inside me rises up to condemn me for not *getting the point* the first time. It's as if one should read and study, grasp the point then move on to discover the next one, as if there is but *one* point to 'get'. This whole inner dialogue is wrought with guilt. On one level, it's okay to re-read a fairytale or a cute children's book to find joy or comfort in the message. But, to re-read a Bible story seems like an admission that you didn't get the point the first time. That *voice* is wrong!

In reality, Bible stories are alive and timeless. If, in humility, I come back to the same old text with expectation and an open heart there is no telling what I will learn from it. The words on the page haven't changed, but *I* have. I'm not the same person as the one who was present at the last reading. The power of Holy words is in their ability to engage the soul of the reader in the present moment and re-create themselves into a new and necessary *moment of truth.* Some of my favorite children's stories have made a powerful impact on my heart. But, ultimately the well-worn 'velveteen rabbit' isn't as mysteriously dynamic as the *lost son* or the *woman caught in adultery*'. It's like God occasionally blesses me with *holy dementia* as I re-engage a character in a New Testament parable again *for the first time.*

Sometimes I rub shoulders with an event or a person or a place that triggers a memory of a deep 'truth' that I have come to believe (or at least have put significant energy into trying to understand). It's like God puts his highlighter through some of his holy text and says, *"Bill, pay attention to what's being played out before you. This is a picture where my Scripture, my words, are being acted out in living color. Watch and learn"*. His words make much more sense to me when I can *see* them being lived out through the flesh and blood of broken faithful people. The following are a few examples of where holy verses have come to life right before my eyes.

Rwanda, East Africa

For I was hungry and you gave me something to eat, I was thirsty and you gave me something to drink, I was a stranger and you invited me in, I needed clothes and you clothed me, I was sick and you looked after me, I was in prison and you came to visit me. Matthew 25:35-36

A letter I wrote after being at a memorial service for our friend Sally…

March 2, 1999 *An open letter to the "Diet Club"*

My dear friends,

I cannot begin to express how much your service to, and love for Sally, has touched me. It is an honor to be called your friend. As 1 sat at the memorial service I realized that I was unworthy to be seated among you, the 'diet club'. I should have moved to a *regular* seat away from that holy ground. Isn't God amazing? And doesn't He have a sense of humor? To think that while you all were developing a 'club', an excuse for some laughs, he was creating something holy.

Calling the gatherings themselves *holy* may be a bit of a stretch but each of you at different times and in numerous ways were holy little tools in the hands of a mighty God. Your *menu* changed slowly, subtly, from a tray of food into a banquet table of love. You delivered life to Sally, one spoonful at a time. Is there a more holy work? Is this not what Jesus himself said was one of the qualities of those whom he allowed to enter his Kingdom? .... "*I was sick and you looked after me...*" My friends, as you served Sally *you* were being Jesus to her. Is there a more important labor on this side of Heaven? The ultimate privilege of it all is that as you cared for Sally you were caring for Jesus *himself*. Isn't it amazing to know that the hand of God is attached to the arms of sinners?

Be at peace, my friends, knowing that you, like Sally, will someday be overwhelmed by Jesus when he shows up again to serve you disguised as the "diet club".

I am blessed to be your friend and fellow sinner.

Maui, Hawaii

What shall we say, then? Shall we go on sinning so that grace may increase?
Romans 6:1

I have been in and around the mission of Young Life in one form or another since my high school days. My friend, Marty, played a significant role in helping develop a Young Life property, Lost Canyon, in Williams, Arizona.[2] There was a point in the spring of 2005 when funding for the last phase of the project dried up. The dust literally settled as the whole thing came to a screeching halt. Marty sent a note to some friends and asked for discernment and wisdom. I am guessing he received some thoughtful words of encouragement, he also received a significant financial gift that re-started the bulldozers and buoyed his spirits. This provision provided me a moment of reflection and I passed along the following words to my friend.

May2, 2005

Marty,

I am re-looking at *Wild at Heart* with a group of guys and came across a classic Eldredge line that seemed to fit your Lost Canyon journey:[3]

"Do you know why God loves writing (or creating: in the case of Lost Canyon) such incredible stories? Because he loves to come through. He loves to show us that he has what it takes."[3]

The Lost Canyon story is filled with countless miracles and 11th hour provisions. And this last huge gift has brought us to tears... Imagine how large our Father's grin was and is as we heard the news. He must really enjoy it each time he has come through in our lives. Lord knows, we've given him plenty of opportunity to do so. Our journeys have often been on the edge. We've been faithful and we've been cowards. We have taken great risks and we have been much too calculated. We have turned our back on Satan's offerings and we have sinned boldly. I am not sure what brings the Lord most joy: When we fall off of the 'edge' and need to be rescued, or when we faithfully say no to an enticing option? I guess Paul struggled with that when he contemplated sinning more so that God might pour out more grace. I am sure our Father does not want us to sin just so he can come through. But, I do find great comfort in knowing his grace is sufficient.

We have taken him on an incredible roller-coaster ride. And the more I think about it the more I have this picture of him not in the back seat holding on and wishing we had made different choices, I see him in the front seat of the first car, hands in the air with a big grin on his face screaming at the top of his lungs, "Hold on tight boys, you ain't seen nothing yet."

Let us continue to enjoy the ride in the name of Jesus.

Early in the morning, Jesus stood on the shore, but the disciples did not realize that it was Jesus. He called out to them, "Friends, haven't you any fish?" "No," they answered. He said, "Throw your net on the right side of the boat and you will find some." When they did, they were unable to haul the net in because of the large number of fish. John 21:4-6

May 11, 1999

I was reading in John 21 where Jesus enabled his friends to have a huge catch of fish.

Why did he show up that day on the beach? I guess to remind them again, before he left, that he was indeed the Son of God. A miraculous catch was a nice way to get their attention but the way he did it intrigues me.

Why did he tell them to cast their nets on the *other side* of the boat?

Why didn't he just fill up their nets in the water where they lay?

Why didn't he just surround the men with 153 flopping fish in the boat'?

The intriguing question for me is: Why did he make them *work*?

This encounter is teaching me about obedience and humility. There seems to always be a response on our part when we encounter Jesus. He is always willing to give extravagantly when we are willing to be obedient. It's not about *working* for the gift but about *responding* to his voice. I must pray for the humility to drag in my empty net at the end of a long night and throw it out again when a voice, THE voice, from a distant shore tells me to do so.

Sea of Galilee, Palestine

"The kingdom of heaven is like treasure hidden in a field, which someone found and hid; then in his joy he goes and sells all that he has and buys that field. Matthew 13:44

A good friend was feeling depressed, without much passion for the Lord or his family. I shared this short note after we had spent a few hours together:

Feb. 17, 1999

Dave,

Thanks for lunch today. It's always a blessing to be with you. I read Matt. 13:44 this afternoon. Why was the man so passionate, so excited to own the treasure? Because he knew its true value! I can't help but believe that if we appreciated *the* treasure as much as this man our passion and zeal for the lesser treasures of life would grow as well. When we sell out for an unworthy treasure or worse, lose our passion for even the less valuable, it's no wonder we get dull and feel empty.

Dave, pray for the Lord to make you be *present* enough to know that you have tripped over *the treasure*. Then you will be able to re-mortgage your life and bask in the joy of its possession.

Let's commit to pray to that end for you, for each other. This would honor God and be a gift to both of our wives.

Read Matt. 13:44 every day until we meet again asking the Lord to rekindle your passion.

The Black River Boys 2012

"Better is one day in your courts, than a thousand elsewhere; I would rather be a doorkeeper in the house of my God then dwell in the tents of the wicked." Psalm 84:10

May 13, 1997

Would I rather spend a day with the Lord over anything else? No! I often dream about lots of things I want to do and dwell on how good they'll be and I forget about residing with the Father. I am dreaming about fishing at the Black River next week thinking that it doesn't get much better than that. I must *believe* it's better than being with the Lord or He would take its place in my thoughts. But maybe being at the Black River with some brothers is like being with the Lord, like being in a mountain cabin with Janny and Casey. Maybe that's why I am yearning for it. Maybe I want to be with Him more than I know and *He* is my motivation. Maybe I just want to escape and the previous two lines are bullshit. Maybe I just love to fish because, after all, the really great disciples were fishermen. Maybe "escaping" once in a while is healthy and holy. My prayer, I guess, is that I would be satisfied as a doorkeeper and yearn to be in the Lord's court, and that next week when I rendezvous with my brothers his *Kingdom will come* and we will fill his courts with praise and fish and *something* holy. I also pray that today I will believe I am in his court and it is holy and just as good as the *court* he is preparing at the Black River.

Phoenix, Arizona

Now there is in Jerusalem near the Sheep Gate a pool, which in Aramaic is called Bethesda and which is surrounded by five covered colonnades. Here a great number of disabled people used to lie—the blind, the lame, the paralyzed. One who was there had been an invalid for thirty-eight years. When Jesus saw him lying there and learned that he had been in this condition for a long time, he asked him, "Do you want to get well?" John 5:2-6

March 11, 2002

I gave a talk to some guys at a golf tournament about the 'need vs. want' thought I have been developing from Jesus' encounter with a man at the pool of Bethesda. Jesus asked do you *want* to get well instead of do you *need* to get well. In the spiritual world, the implication is that we understand *need* but there is hesitancy to *want* because it's potentially too costly. Mark suggested I use a smoker as an object lesson since the 'need' factor to quit is huge but the 'want' is often not. This morning Mark called and offered a new insight. As children we have numerous memories of a parent asking us, "Do you need it or just want it?" As children we just *want* stuff, *need* never plays a role in the decision. *Kids want, adults need.* Kids don't bother themselves with needs. Adults may ignore *wants* after calculating the cost. What does this thought, if any, have to do with the fact Jesus calls us to be like little children if we are to enter the Kingdom of God? There must be something about childlike *wanting*, about being carefree and not getting sidetracked by the potential consequences of our actions that pleases Jesus.

Charlotte at Lake Gilmore, Wisconsin

"...and it takes everything as it comes, like a child, who, being in a chronic state of wonder, is surprised by nothing."[4]

He called a little child to him, and placed the child among them. And he said: "Truly I tell you, unless you change and become like little children, you will never enter the kingdom of heaven. Matthew 18:2-3

Is there something to the childlike 'wanting' part of being young? Is it holy to stay in the 'want' stage of life and not get stuck in the world of 'need'? The idea of maturing, getting wiser and smarter, might be overrated in a theological sense. When we over think some things, simple things, we get confused and can actually change the *flavor* of the concept altogether. Maybe the idea of grace or trust or faith is *vanilla*. But maybe we want it to be *marble fudge*. I mean after all, vanilla is boring. Sophisticated people need a fancier dessert; something more glamorous and complex. But the child has only tasted vanilla. Sure he loves it. But that's only because he has never tried Rocky Road or Turtle Fudge or Pralines and Cream. Once he gets his taste buds around one of those more complex flavors, he'll never look back. He will *need* a flavor from now on. No longer will he want just plain old vanilla ice cream. Now he will be selective, contemplative, discerning because he knows he has *choice*.

That's it! *Choice* is the key to the problem. Is this the real problem with humans? Did it all start with a choice? Adam and Eve were happy/content until the possibility of choice was before them. Should we withhold choice in order to manufacture allegiance-faith? No. Without choice there can be no faith. We *want* flavors to choose from. But we only *need* vanilla. Yet we need choices in order to know what we want. I believe I need to know there are options in order to *want* in a healthy sense.

"Come to me, all you who are weary and burdened, and I will give you rest. Take my yoke upon you and learn from me, for I am gentle and humble in heart, and you will find rest for your souls. For my yoke is easy and my burden is light."
Matthew 11:28-30

March 20, 2008

I just finished reading *The Body Broken* by Robert Benson. I am thinking this morning about the verb 'is'. His title: *The Body Broken*, really resonates with me for some reason. I think because 'is' isn't in the title. The verb's addition would almost imply judgment or condemnation. By leaving it *out* of the title the words draw more listeners *in. The Body Broken* communicates a sort of sympathy and care. 'Is' would make it a statement of fact, but the lack of 'is' makes me imagine a sweet parental moment of comfort. Yes, something, some part of the body, *is* damaged, injured or broken. But, by not saying 'is' the reader isn't insulted and won't feel like responding with a sarcastic, "You think?" But instead, those sweet and kind parental arms of comfort and concern just reach out and say "come"... Jesus said, "Come unto me you who are heavy laden." What a wonderful moment where *fact* and *love* intersect. The verb here doesn't seem to imply judgment, it simply opens the door to comfort.

When he had finished speaking, he said to Simon, "Put out into deep water, and let down the nets for a catch." Simon answered, "Master, we've worked hard all night and haven't caught anything. But because you say so, I will let down the nets." When they had done so, they caught such a large number of fish that their nets began to break. So they signaled their partners in the other boat to come and help them, and they came and filled both boats so full that they began to sink. When Simon Peter saw this, he fell at Jesus' knees and said, "Go away from me, Lord; I am a sinful man!" For he and all his companions were astonished at the catch of fish they had taken, and so were James and John, the sons of Zebedee, Simon's partners. Then Jesus said to Simon, "Don't be afraid; from now on you will fish for people." So they pulled their boats up on shore, left everything and followed him. *Luke 5:4-11*

April 30, 2010

I read in Luke's gospel that the new disciples "left everything" to follow Jesus. They were clearly astonished at the miraculous catch of fish, the huge provision that would've brought in a serious amount of money. I can imagine them thinking things like; 'this could pay all my debt/loans', 'I could get some much-needed new equipment', 'I can finally go on a decent vacation', 'Let's figure out how many times a week we should ask Jesus to do this so we can have an easier life...' Why was it so easy, natural, for them to walk away into the unknown? Did they sell the fish and their boats to finance their mission? Were they prepared for this moment by a deep spiritual walk and connection with God? If they were fishermen because they did not make the grade in Rabbi school (as Rob Bell has suggested [5]) then I wonder if some thought, "well I finally got a shot with *somebody*, so I'm going for it." Would a desire for religious training, growth in God, be strong enough to cause an uneducated Rabbi wannabe to walk away from his vocation? Are we in the West so comforted and seduced by material things that it is impossible to think of exchanging *stuff* for *Spirit*? Lord, help me stay sensitive to your calling, be open to astonishment, so wealth does not cloud my senses.

Since my youth, God, you have taught me,
and to this day I declare your marvelous deeds.
Even when I am old and gray,do not forsake me, my God,
till I declare your power to the next generation,
your mighty acts to all who are to come. Psalm 71:17-18

October 31, 2010

I'm reading Psalm 71 this morning and have these thoughts;

David is considered "a man after God's own heart". An A-team guy for sure, probably the captain of the team. His words are often pleading for God to save him, often from a foe. But today I'm getting the sense of insecurity and uncertainty in his voice. He begs the Lord to save him and not let him be put to shame. How many times does he have to beg for rescue and get rescued before he has peace and confidence in his position with God?

At the end of this Psalm he reminds God of his faithfulness in his youth and asks God not to forsake him when he is old. This regular dichotomous thinking: confidence/thankfulness vs. concern and uncertainty confuses me. It reminds me of a Catholic funeral that is saturated with the confidence of salvation for the deceased and his/her place with the Saints in heaven, but still sets aside two or three times to pray for the dead friend and plead with God to have mercy and receive him/her.

I guess I don't sense a deep confidence in David that all the grace God has bestowed upon him along the way has convinced him that he will receive grace in the end. Maybe David is praised by God as *a man after his own heart* because for a person so steeped in a *works* religion this is the best He can hope for. Maybe, occasional outbursts of faith and rejoicing book ended by 'please don't leave me' is all one can expect considering where David started from.

That is *amazing grace* in my eyes.

As I continue to pass judgment on the *righteous* that surround me (an unfortunate bad habit), I must consider from whence they started…

Oh Lord, forgive my judgmental spirit, and thank you for your love and grace in *my* story, life and heart. I am grateful to be your beloved child.

The Wave, Vermillion Cliffs Wilderness, Arizona

Jesus told them another parable: "The kingdom of heaven is like a man who sowed good seed in his field. But while everyone was sleeping, his enemy came and sowed weeds among the wheat, and went away. When the wheat sprouted and formed heads, then the weeds also appeared. "The owner's servants came to him and said, 'Sir, didn't you sow good seed in your field? Where then did the weeds come from?' "'An enemy did this,' he replied. "The servants asked him, 'Do you want us to go and pull them up?'

"'No,' he answered, 'because while you are pulling the weeds, you may uproot the wheat with them. Let both grow together until the harvest. At that time I will tell the harvesters: First collect the weeds and tie them in bundles to be burned; then gather the wheat and bring it into my barn.'" Matthew 13:24-30

June 14, 2011

I have been noticing gardening metaphors lately they seem to be everywhere. Maybe I am sensitive to them because of my focus on my new garden. Allowing the weeds to grow up with the wheat in Matthew 13 has always been a challenging concept. Even though it's a known fact the weeds were sown by the enemy, Jesus says that because some are too close to the wheat they will have to stay until the harvest to protect the wheat from being pulled up. Today I am thinking that in my garden I planted some seeds too close together. One squash plant flourished and produced some good fruit while several of them sprouted but didn't get very big. I think they were too close together and crowded each other out, seems like I should've thinned them out. I could've transplanted some, not just pulled them out. I got to thinking about a *seed* being a church. Could it be possible to have so many churches so close together that they couldn't thrive and produce fruit? And who would hear the Holy Farmer if he said, "let's dig this one up and replant it over there." Would He intercede at this point or would He say, "no, leave them altogether, while you try to dig one up you may uproot the one next to it. Leave them until harvest time when I call the reapers to collect the wheat. Some of them will have a large amount of grain, some a little. What about the *plants* with little or no grain? Why didn't He transplant them early on or just pull them out? I guess because His economy is different than mine. He doesn't judge like I do. If he had enough faith in some wheat to live side-by-side with evil weeds I guess he figures some strong wheat can handle some weak plants next door.

Flagstaff, Arizona

"When you enter a town and are welcomed, eat what is offered to you. Heal the sick who are there and tell them, 'The kingdom of God has come near to you.' But when you enter a town and are not welcomed, go into its streets and say, 'Even the dust of your town we wipe from our feet as a warning to you. Yet be sure of this: The kingdom of God has come near." Luke 10:8-11

September 14, 2011 The Refreshment of Innocence

I am so encouraged by the way Daniel Chelgren keeps helping me learn about Scripture. Yesterday we were reading and talking about Luke chapter 10 when he had another one of his profound observations: an observation that could only come from an uncluttered and unadulterated heart; where innocence prevails and simple questions are free to surface. This is the polar opposite of the tension that surrounds the classic "*squirrel–Jesus*" dilemma for kids who have endured too many Sunday school classes. Here is the context: Jesus sends out 72 of his friends in pairs to tell everyone that the Kingdom of God is near. He instructs them to enter the towns that welcome them and stay. In the places where they are not invited in, they are to shake the dust off their feet and let those people know their future is bleak because of their disinterest. I know this story, I have read it a lot. And I have mined the *jewels* from its depths and bagged them up, along with a few lingering questions. I was poised and ready to reveal my wisdom (the contents of my *bag*) as we sat there in that coffee shop. Then Daniel drops a *jewel* of his own on the table. His doesn't come from a well-worn pouch deep in his pocket. His shiny fresh stone tumbles out of a fuzzy velvet ring box and it shames my dusty offering. He wonders, "What if the first guy Jesus' friends meet as they enter a town tells them to 'go away we're not interested', and he's not speaking for the townspeople but only himself? Maybe he's had a bad day or is an anti-Semitic God hater or something. Then the whole town would be screwed." This is a brand-new thought to me. These kinds of gems regularly come out of a fresh place in Daniel's spirit. We had a good time thinking about the practical part of this question: maybe Jesus' instruction to not greet anyone on the road meant they were to always go directly to the town center and ask the Mayor or something. Finding an answer to his question is simply academic. What is so energizing to me is the freshness and innocence of the very *thought*.

It's these moments that make me ask myself if I am too learned to be refreshed by the Gospel stories I have read and studied for years. Have I bagged up all my gems and labeled them by book and chapter so I can dust them off from time to time to use to enlighten a hungry disciple who comes across my path? Am I too theologically refined to be spiritually refreshed by God's words? All too often the answer is; yes. I am so grateful that Daniel spends time with me. He, and other young guys like him, are God's way, his voice if you will, to expose the cracks in the theological and ego-bolstered walls of my intellect. Their insights refresh my spirit and humble my soul. I pray the Lord keeps allowing guys like Daniel to speak into my life.

When John, who was in prison, heard about the deeds of the Messiah, he sent his disciples to ask him, "Are you the one who is to come, or should we expect someone else?" Jesus replied, "Go back and report to John what you hear and see: The blind receive sight, the lame walk, those who have leprosy are cleansed, the deaf hear, the dead are raised, and the good news is proclaimed to the poor. Matthew 11:2-5

July 15, 2003

I am doing a little study on 'how Jesus communicated' with my friend Rick. In Matthew 11 John asks his disciples to go ask Jesus if he is the one who was to come or should we wait for someone else Jesus could've answered, "Which *one* are you talking about?" Or, he could've said, "Yes, I'm the *one*." Or, he could've said, "John should know since he's my cousin." But instead, he said go report to John what you see: the blind see, the deaf hear, the lame walk, the dead are raised to life, good news is preached to the poor. It seems to me he was saying, "Stick around and see for yourselves if I am the *one*." Watch and you will know the answer. If he had just come out and said, "Yes, I'm the one", they probably would not have believed him. So, instead he told them to watch and see. Am I so patient when people ask me, "is Jesus really God—is he the one—the answer—the way?" Can I confidently reply 'wait and see' instead of feeling like I have to say *yes*? Am I okay with leaving all the *conversion* power in God's hands? I feel compelled to prove with my words, intellect and power of persuasion, that Jesus is the *one*. I am sure I should be more at peace with 'come and see, check him out'. This is a challenge.

Torroweap, North Rim of Grand Canyon, Arizona

5

GOD; HIS CHARACTER

September 10, 2006

Our 28th anniversary was yesterday. We are in Lincoln City, Oregon listening to the Lord shouting at us from the crashing waves while awaiting his next sunset. Since God lives outside of time I think I will venture to say this will be his first sunset – his *only* one. If, indeed, he lives only in the *now*—this moment, I wonder if he is like a man with Alzheimer's who meets his best friend or wife anew every day? Does God create a sunset and get all giddy and excited over it? Does he live in such a chronic state of *presence* that everything excites him each time he makes it? There are miles of very old logs along the beach here. Does he remember when they were tall and strong and green? Or does he not burden himself with the pain of memory and just enjoy watching people use them today for a wind block or to warm themselves by a fire tonight? He calls us not to worry about tomorrow, maybe because there is no tomorrow for him, and no yesterday as well. Maybe, he is only trying to get us to become more like him and live right now, love right now, BE right now.

Lord let's rejoice right now, together, while you paint the world a masterpiece of yellows, oranges and blues. Ah, forget the rest of the world, maybe this one is just for me. I love you!

Lincoln City, Oregon

September 22, 2010

Kristy Matock gave birth to her stillborn daughter last Saturday. Kinsley was 34 weeks 5 days 'old' – the same age Casey was when she was born, alive. Kristi and RC want Psalm 139:13-16 to be read at the Memorial. They didn't know that those very verses were the ones we chose to put on Casey's isolette in her early days at the hospital. God 'formed' Kinsley and Casey and 'numbered' both of their days. The *forming* part is easy to credit God for it's miraculous and intricate and special. The numbering seems arbitrary and risky. Why did Casey get more days than Kinsley? The question itself assumes or implies *more* is *better*. More days with a child, a friend, a mate is always better – isn't it? We dream of a future for our children; adventures, experiences, special events… Oh, those visions we have of the future filled with the best ideas our limited minds can conjure up. We 'number' our children's days and then we hope or often expect God to read from *our* script. And when the story doesn't turn out the way we wanted it to or hoped it would we respond with anger, sadness, bewilderment and shock—

Wasn't God listening? Didn't he get the memo? Does God know what's happening? This is what I *know*—what I believe: God loves all of his children because he formed them all. Outside of our limited linear world, He exists. He is 'knitting' in a timeless place that has no beginning and no end. Where nothing gets broken and there are no interruptions or mistakes or surprises or pain or disappointment. Kinsley is in that place where it is perfect, where love prevails in the arms of her Father, her Maker. She is *really* living, not bound by time. But, on this side of eternity we are bound, limited. Our finite minds cannot comprehend the bigger picture. We are bound by our plans and dreams and hopes for ourselves, for Kinsley.

I choose today to surrender to God and trust him. I want to give up control and do *what* I can *while* I can. Jesus gave three commands while he walked the earth: love God, love your neighbor and love one another. There are no promises or rewards attached to these commands. They don't require analysis or study, love is immeasurable. We are called to love from the One who is called Love. Today I want to be satisfied with my Father's love—may it be enough. May I be at peace as I do my best to love the *living* before they and or I have opportunity to be with Love for eternity.

West Fork of Oak Creek, Arizona

December 27, 2011

I read an article about a local science teacher who has made a great impact around here. His mantra is "everything is science". It made me think of Brad sharing with me that “everything is art”. I got to thinking that I could argue *everything is God* – theology. Or the mathematician saying *everything is math*. What if ‘everything’ was *all* of these things in equal parts? (I mean except for God). Clearly, all created ‘things’—‘ology’s’ were formed by God and combined to work perfectly together, hand-in-hand. If one was taken away then all the others would be incomplete and the world would be *out of balance*. Maybe, when you buy into your *thing* to such a degree that it nudges out or overshadows the *other things,* you are doing the whole cosmos a disservice and ultimately are messing up God's design.

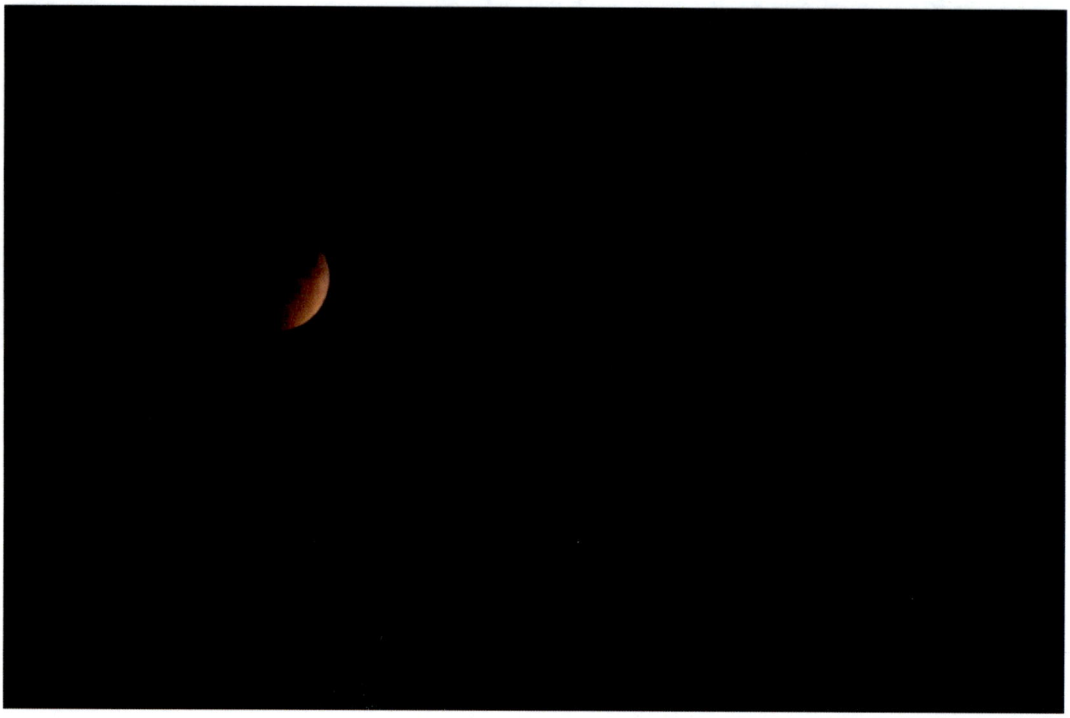

Lunar Eclipse, Phoenix, September 27, 2015

March 19, 2013

I have been in discussions lately with friends discussing the idea of following the wrong Jesus. When my faith journey was beginning to gain traction in high school I am sure I received a lot of encouragement to fear God. I recall being told by someone to picture Jesus in the backseat of the car when I was out on a date. The thought of seeing his face in the rearview mirror glaring at me helped me keep my hands to myself, mostly. I tried hard not to disappoint him or make him mad. Eventually I grew to believe this is not the correct image of the *real* Jesus. He was not waiting to get mad at me or keep score of my missteps along the way. *The* real Jesus loves me all the time, even when I disregard him. I slowly came to understand that the real motivation to keep my hands to myself should come as a response to his amazing love for me not from fear of what's recorded on some holy scorecard. The first time I experienced this kind of real love I was just a kid and the encounter became an important building block in my spiritual formation.

A very nice older lady in our neighborhood was a leader of a Good News Club. Kids in the neighborhood would go to her house every week and listen to a Bible story and have treats. If one of the kids brought a new friend to a meeting he/she would be allowed to reach into a bag for a special prize. That's how I ended up darkening the door of her house for my first and only Good News Club. I do not recall anything about that meeting other than feeling uncomfortable and *used* by my friend. The *formative moment* came when a friend and I were riding our bikes down the alley behind the Good News Club house a few weeks later. We heard a group of kids playing some sort of a game in the yard. We stopped and picked a dozen low hanging grape fruit *bombs* and unloaded them into the yard, scattering the crowd. Mrs. *Good News,* with her very kind and sweet voice said, "Thank you boys for the nice grapefruit." She thanked us for raining fruit down on her innocent flock, and in my spirit I knew she *meant* it. I was so ashamed at that moment. I was expecting harsh words of condemnation, but instead, I felt her compassion and love for a broken kid. *She thanked me*! She showed me love when I deserved to be yelled at and forced to clean up the mess. She was being the *real* Jesus to me. Her witness to me and those kids was powerful. This is my first memory of experiencing undeserved love.

West Fork of Oak Creek, Arizona

March 25, 2010

The Lord has awakened me this morning early to cast light upon my anxiety. As I look now at my Journal I can see my last entry—the last time I felt compelled to write was motivated by the same fearful place I am in now. Pathetic. But, this morning He is speaking to my heart with His truth. It is embarrassing and shameful to be depressed and disconnected by fear, especially the fear of lack of money. But, He is filling my mind with words of truth:

Trust in the Lord and he will make straight your paths.

Jesus has a place prepared for me.

Rejoice! Your name is in the book of life.

The peace of God which surpasses all understanding is mine in Jesus.

For God so loved me he allowed his son to die for me.

Do not worry about tomorrow or about today – food clothes etc. I'll take care of you.

Come unto me all you who are heavy laden – my burden is light.

Cast all of your cares upon me.

I can't shake the picture of Jesus on the shore of the Sea of Galilee cooking breakfast for his friends while they are struggling to haul in his miraculous provision of fish, and fully clothed Peter flailing and splashing his way to the shore.

Lord, is this a word for me now? Am I out 'fishing' instead of *being* with you? Is it me in that boat just 30 days after the greatest moment of my life; seeing the resurrected Jesus? Am I beginning to think, "well it's been fun, but now it's back to fishing again"? So, on this *bad day of fishing* you are standing on the *shore*, the edge of my bed, all night saying, *"Come to the shore where my warm and nourishing provision awaits you? And, by the way, go ahead and haul in those fish. There are plenty more, the lake is full of them. Cast your net in faith but remember, I am always here on the shore with a warm fire and nourishment. Lo, I am with you always... You will get recharged by standing at the campfire with me. Come and stand with me. I love you."*

6

SEEING JESUS IN PEOPLE

June 20, 1994   Learning about joy at Woodleaf

The Capernaum kids did the ropes course today. The highlight of my time here to date. The able-bodied ones actually went through the whole course. They were beside themselves with excitement, the single greatest physical accomplishment of their lives. The kids in wheelchairs got hoisted up out of their chair to the bar so they could reach out and touch it. The faces of those kids as they arrive back into their chairs were so alive, so overwhelmed with joy. Part of me wants to think they must feel so disappointed after their taste of freedom, that to sit back in the confines of the chair would be depressing. But the moment of ecstasy is so large that that thought never seems to cross their mind. To see these kids sit at the top of the platform and be so afraid to jump – then finally go off is powerful. Not because of what they did, but because of their capacity to not remember the fear of failure only the victory. Would to God that I might be able to enjoy the victory for the simplicity of it—for what it is without judging myself or comparing and measuring it. I wonder what it feels like to be saturated with joy over a victory instead of inundated with analysis. This is a worthy prayer request. Thank you Lord for the privilege of seeing the kind of heart you would have me possess.

October 28, 1998   The life behind the words

Jan and I went to hear Jackson Browne in concert last night. He was solo, acoustic. I really enjoyed it. His songs are similar, mellow, thinking, they have meaning, message. He is trying to say something in his music in a style that I appreciate. I found myself wanting to meet him and hear his story. Not because he's 'famous'. But because of something inside me that wants to reach out to people with contemplative hearts. I really enjoyed listening last night. Was it about nostalgia, old high school memories brought back? I don't think so. Something happens to me when I hear music that has words that resonate with hope, despair, fear and joy. For me, the story doesn't have to be about Jesus but, the words have to compel me to think and care, to identify with them. Maybe that's it. I identify with words and thoughts like that. Maybe I have spent a lot of forgotten time as a kid and young adult feeling the feelings that I have when I listen to David Wilcox or James Taylor sing, or when I read Frederick Buechner or C.S. Lewis. I like what it does for my soul. I hope this liking is growing out of health and not denial or depression. I want to meet these people. But, really I guess I want to meet the reason they write or sing…

March 22, 2000   Via Maria del Mal, Santa Cruz, California

This Catholic retreat center is right on the Monterey Bay. What an amazing place to sit—on the beach watching and listening to the Lord. I saw some piece of broken bottles on the beach as I walked this morning. I thought I would pick them up on my way back. Some negligent person threw their beer bottle into the surf. Now some innocent person is going to get cut. I stopped at a place where the sand gave way to gravel. In the gravel were colored 'rocks', actually, 'sand glass', old pieces of glass worn smooth from many years of being in the Lord's tumbler. They were a great contribution to the mosaic of pebbles. I picked some up: green, brown, clear and a cobalt blue one. They're smooth, worn, almost soft to look at. They mix in well with the rocks. They must've been sharp and dangerous when they were newly discarded. They may have a history of causing pain but now they are harmless and beautiful. They needed to be broken and discarded into their ocean prison in order to mature into the current state they are in. Is this a picture of how all men naturally mature or is it a picture of the path only a few go down? What if the bottle was never broken and it was well taken care of; used, dutifully recycled so as to not harm the environment and then became reincarnated (reconstituted) into a road or something? Who makes that choice? Can a beer bottle stay just a bottle all of its life and get refilled over and over again or at least get emptied and filled with flowers that will wilt and die? I know today that being hurled off of a cliff onto the rocky beach is the road to a humbling, rhythmic journey that will lead, in time, to a smooth, colorful existence among many plain ordinary rocks. I think I would rather be found as a worn, former piece of trash, on a beach than a non-descript spec of composite getting driven on by ungrateful passersby. Lord, today when I see a jagged discarded person I will thank you in advance for what you are going to tumble him or her into. May he/she become a bright blue light on a gray gravelly beach. Thank you that I am believing more and more each day that you are pushing me closer and closer to the beach. I love you.

Punta Mita, Mexico

August 6, 2000 On a road trip to northern Montana with Stu

We drove 22 hours, 1,356 miles to his sister's cabin in Fortine (north of Whitefish). It's brand-new and beautiful. We pulled in early this morning to Whitefish and stumbled upon folks in the park who were preparing for a special worship service: The 75th anniversary of Christ Lutheran Church of Whitefish. From the cab of our truck we kept our eyes peeled, looking for Jesus. He showed up in need of $.80. Guess he got drunk at a concert last night and got separated from his family. He smelled of alcohol and was very friendly. Then, two different ladies gave us the judgment eye, like we should leave this holy place. We didn't look too good having just slept in the cab of our truck. I wonder; did Jesus ever look like us after a long night of travel? Were there people like these ladies, back in the day that missed Him because his *cover* was incorrect? Then we got another Jesus sighting. This time he showed up as an older man. He was talking with three skater kids encouraging them to stay for the music, worship and picnic. He gave one of them a name tag. It seems that Jesus still knows all our names – just like Zacchaeus. This evening I experienced Him in the mountains to the north, in his forest, in the sounds from the wind and in Stu's guitar. Where will we meet Him next? I love Jesus in all of his disguises. I hope I have eyes to see tomorrow.

Jackson Hole, Wyoming

January 8, 2001   Sunday morning, Church at the Neighborhood Center

A woman during the prayer requests (translated) asked God to forgive her because she was unable to give her daughter much of a celebration for her birthday, only love. Casey and I both wept (Jan wasn't there). Can you imagine asking God to forgive you for *only* being able to give love with no celebration to your child on his or her birthday? In the world where I walk, the birthday *celebration* often outweighs the *love*. Most of the time we should be asking God for forgiveness because our *celebrations* overshadow our love. That wonderful lady ministered to me, to my twisted heart. I will write her a letter—

Lake Gilmore, Wisconsin

October 21, 2009

These thoughts came after I returned from California helping Tony and Carolyn through their first days after Morgan's accident.

I was able to feel compassion—real deep compassion the first morning we arrived at the hospital. I don't experience real compassion very often. I'm not sure why, must be a form of self-protection or just deep-seated selfishness. But, as I looked at Morgan all 'connected' to life support and not knowing if she would live or die, I saw my own daughter, Casey. Only then did I *know* how Tony and Carolyn felt. I experienced their fear, pain and angst. I *knew* compassion. It's foreign, but holy, territory. I knew also that I never want to stand alone over my child or wife in a similar situation. As always, these moments of service I find myself in are mostly to teach me about *me* and consequently deepen my journey with my God.

December 3, 2009

I went to see my doctor yesterday to pick up a prescription. He tells me he has a *treat* for me and invites a woman into our room and closes the door. He proceeds to describe to her my vocation in a very encouraging way–very affirming. He said I am an evangelist—the "highest calling". Then they laid hands on me and he prayed for me to have favor and power for healing and that he was amazed at the courage it took to do what I do. It was so affirming. Then she asked if she could share something with me as he left the room. She told me that God got ahold of her about three years ago and has called her to *prayer*. So she prays for people and often gets 'pictures' to share with them. She got a *picture* for me and asked permission to share it. She saw me with lots of rings in my hands and she asked the Lord what the rings meant. ***He said that I put rings on people's fingers and restore their relationship with God.*** Wow! I affirmed those words, thanked her, and walked away from the best doctor visit of my life.

7

SOUNDS

December 31, 2001 Canaan in the Desert[6]

I hear the calling of quail and it makes me think of evening in the desert. I wonder why the quail are calling at midday? Then I realized the quail probably call all day long. I've only stopped long enough in the evenings of my past to hear them. The quail are not doing the extraordinary – they are behaving quite normally.

Where do I hear Jesus? When? Is he not speaking always? I must remember to not make a law or ritual out of my wonderful communion with him and remember that he is calling even at midday.

When you sit in the desert and wait, watch and listen, you realize how much is taking place all around you. A cactus wren is catching bees for lunch, no hands, just a beak. The palm leaves rub against each other and make a sort of clumsy sound. The fountain is splashing relentlessly. A hummingbird has parked itself on the top of a Palo Verde tree. There are empty husks of small seeds on a flat rock – some creature must've dined here today. The sounds of cars on 40th Street seem to violate this space. I hear the cooing of a dove: one of the finest sounds to my ear. Now, quail are calling out. The dove sound soothes me—why? As a child, I must've had countless evening hours in the desert hearing these songs. Quail walk by, a thrasher is calling out from a prickly pear. A cactus wren is getting noisy and walking all around me, maybe it's been stung by a bee.

Murchinson Falls National Park, Uganda, East Africa

February 11, 2004 Canaan in the Desert

I love these sights and sounds. Are these feelings born out of a peaceful memory, nostalgia, or sad and lonely times in the desert as a child? I do not know, but it really doesn't matter. I like this moment: this ‘connection’. I'm aware that I do not hear the sounds from the road unless I try, unless I make myself aware of them. Yet, they are mostly constant. I select what I want to hear. This could be a gift or a curse. It is nice that I might not hear sounds that invade and irritate, but what about sounds that heal and soothe, or sounds that teach or expose? How much control do I really have? Is it all subconscious or by design? What words might I be missing right now? Is the Lord speaking at all times to me? I'm quite sure I do not incline my ear to hear His words enough. I love the soothing sights and sounds of the desert. But, I only see and hear when I come and listen. Is it not the same with God? I must come out and listen…

July 24, 2007 Beyond Malibu base camp

The waterfall that sings to base camp is such a blessing. A constant reminder of God's provision of water, his power, and how he uses sound to sooth and massage one's psyche. As I sit here I realize that the sound changes depending on my position in my chair. Leaning forward there's a different pitch than when I lean back. I'm not sure if it is clearer or better in one place or not, but different for sure. The falling of the water doesn't change but my position does. This seems like a metaphor of the Lords voice to me. He does not change his tone or volume, what changes is me, my position. I can hear it no matter where I am sitting but it may be clearer in certain positions. I must put myself in a position to hear the clearest words he has for me. I also realize that although the waterfall is constant and relentless, most of the day I don't hear it. Its voice must just blend in or fade away and get lost amidst the other sounds and distractions of the day. If I stop and listen, step away from the other 'sounds' that are bombarding me, I will hear the thundering velvet sound of my father's voice. Not necessarily trying to tell me something specific, but being a soothing reminder of his constant presence and flow in my life.

Stop and listen. Calm my heart. Thank Him for his provision and hydration. Then try and be part of that flow into those around me who have not heard this sound, who aren't aware of his voice.

Beyond Malibu Base Camp, Princess Louisa Inlet, British Columbia, Canada

July 12, 2011 Two solitude days at a cabin in Christopher Creek

Drove up to the Rim to worship (walk in the Lord's creation). I asked Him for a word, a message. It came: "take time and stop to look and listen." Each time I stopped during my hike along the trail and waited I saw something that would have been missed or lost had I just kept walking: the glistening water droplets on the oak tree leaves; the shadow of the crow catching thermals at the edge of the Rim; the hummingbird coming to feed from the red trumpet like flowers on a weed that I stopped to photograph; the beautiful butterfly that arrived to pollinate the flower of a thistle that I had stopped to photograph; two small tin hearts nailed to the roots of a Ponderosa pine memorializing relatives who had died; the wind in the trees.

These are all sounds and sights I would have walked by or ignored had I not stopped. This is a strong word from the Lord, a reminder, a metaphor, for me to stop in my daily life and wait on the Lord. He has special sights and sounds—words; that can only be experienced in stillness. This holy moment and amazing photos are rich reminders for me to slow down and pay attention…

Hummingbird at the Mogollon Rim, Arizona

Thoughts from the deck:

*Everything is in perfect focus with my glasses on. Why don't I wear them all the time? I appreciate the trees, mountains and sky so much, even better when they are perfectly clear. I should have prescription sunglasses and see things the way they are. But, I don't because I'm cheap. I wonder if I endure other out of focus things because I am cheap.

*Birds are louder when you listen to them. I wonder how many voices around me would be louder if I listened to them.

*As Casey (and Brad) makes an impact in the world she won't be compared to or identified by me and Jan because of her last name. Her accomplishments will not reflect on us (on the surface) and it will be our little secret. Pretty cool.

*Cheese, bread, wine, a cigar, 75° and James Taylor are a fine way to commune with Jesus!

View from the deck

8

ME

August 21, 2011

These thoughts are taken from Ken Gire.[7] They were written during a time of introspection as Ken was contemplating his new vocation of writing about Jesus.

You read books to borrow therefrom the force to stimulate your activity... but I read books searching for the man who has written them. Vincent van Gogh in a letter to his sister.

Wondering where to start, I began looking over my shelves of theological books. It was there I made an unsettling discovery. I had more books on Greek grammar than I did on the life of Christ.

It was incriminating to realize that He who had given so much occupied so small a shelf in my life. In the quiet courtroom of my heart, I was suddenly the defendant, suddenly the one put on the witness stand and called to give an account of my life. The questions were indicting.

What had I been doing in seminary? Had I been learning how to live my life, or had I been learning how to use my gift?

What had I been pursuing those four years? A Savior, or simply a skill?

If the truth, the whole truth, and nothing but the truth were known, what verdict would be handed down to us, to you and me, about what we've been pursuing for so much of our lives, and why?

Had I been reading the Bible the way van Gogh's sister read books, "*to borrow therefrom the force to stimulate my activity*?"

Had I read it, searching for principles, to make my life in some way more successful?

Had I read it, searching for promises, to make my life in some way more safe?

Had I read it, searching for proof texts, to give certainty to my own faith or make it more defensible to others?

Had I read it, searching for preaching material, because that was my job?

Had I read it, searching for power, for whatever reason?

Or had I read it, as van Gogh had read his books, *searching for the man who has written them?*

Wow, much of Gire's sentiments could be my own. During my seminary education (that spanned 12 years) I found myself making notes in the margins of books and highlighting key points in the lecture notes that I thought would be worthy material to teach or share with those under my leadership. I would write: *leaders, kids, committee, staff* in the margins with arrows pointing to the statement or insight that I could use to teach, train, or encourage a specific group with. I know there is nothing wrong with caring for my friends and colleagues in this deliberate manner. In fact, I am sure most of the time my efforts brought blessing. But after a number of years, I had a moment of revelation that was very similar to the one Ken Gire so eloquently writes about in the words above.

I can remember one day, as I was sifting through lecture notes I had taken from a class (where I had sat under the teaching of one of the many world-class theologians I had the privilege of learning from during my formal educational years) looking for a nugget of truth to use to teach *someone* when God interrupted; and *spoke*. The revelation, the same one Gire is alluding to, hit me. Why it took so long for me to get His message is a mystery to me. What I heard was this: These notes you are taking and these truths you are learning are for *you*. The arrows are pointing to words that *you* must believe. And the highlighted texts are truths *you* must embrace. Being a conduit of wisdom is nice but I want you to teach this stuff to *you*.

Jesus has so much more power when he comes to a heart that has *communed* with him as opposed to one who has just *learned* about him. This moment started me down a new path. It's slow going— (Hey, I took a lot of years of notes before I decided to change course…) But I'm attempting to live this revelation. I still make notes in the margins of books I read. I still glean and journal truths from things I hear. But, now the first stop on this information highway is at the doorstep of my own heart and soul where I am asking Jesus to conform me – my thinking, my being. My hope is that I would stay committed to *communion* before *communication*. That I would never again share great words from a page before those words have had their way with my own soul.

Mogollon Rim, Arizona

June 1, 2010   Lost Canyon assignment

A flood of memories have come roaring back as I sit through meetings, watch buses get greeted by a throng of motivated work crew kids and listen to the sound of campers… My last Young Life assignment was 14 years ago at Oakbridge.

My 1st act as Head Leader was a sweet 'welcome back Bill' from Jesus. One hour into the first day of camp a leader hands me a joint he confiscated from a kid on the bus coming up. Nice touch Jesus, as if I might have forgotten the legacy of Young Life in my own story. Me acting cool at Shadow Valley Ranch in 1970 getting high in the forest, which led to grace, love and acceptance by men and women who had no idea if the love they poured over and into me would ever matter, if it would take. So, I can hear the Lord say 'welcome back Bill.' We're only 60 miles and 40 years down the road from where my cloud of smoke wasn't too much for Spirit led people to see past to a kid who was a work in progress. Lord, continue to grant me eyes that see past the smoke that engulfs many who you guide down my path.

November 18, 2004

I'm in Lebanon. Yesterday we drove north to Harissa where we visited a Maronite worship Center. A statue of Our Lady of Lebanon is the focal point at the top of a hill. We climbed the stairs to the top of the shrine were people pray as they touch the statue and give alms. I got up there and looked out over the country and thought about where I was. I marveled at the historic significance of this very old land and how much the whole world was aware of this small piece of the earth. I prayed for the world. It brought me to tears. For the first time in my life I prayed for the world, I believe I meant it. It seemed like my prayer mattered, like somehow praying for the world from a tower in Lebanon while looking toward Israel (Occupied Palestine) meant more than doing it from my house in Phoenix while staring into the backyard. Whatever the reason, it moved my soul as I prayed for the world from the tower of Our Lady of Lebanon.

October 24, 2012

A photograph is like the written words in my journal. All day long I am being bombarded by images. Some are beautiful, some ordinary. Often, I feel compelled to *capture* one – to save an image that I think is either unique, amazing or special. Journaling is like that for me. Ideas and thoughts are constantly bombarding my brain and psyche. Until I capture one and allow it to come into focus in the form of words on a piece of paper it doesn't have much of a chance to be seen or appreciated. My ADHD condition requires the 'shutter' of my mind to be released in order to capture that beautiful thought and record it in words. Maybe the joy or pleasure I received in taking photographs has a correlation or connection to my ADHD.

November 10, 2012

I just reread my thoughts from October 24, 2012 and feel compelled to expand on the ADHD part. To me, this condition inhibits my ability to focus for a long period of time on a specific thing or idea. It's difficult to allow myself to 'release the shutter' and *capture* the idea, concept, moment. There is clarity and power in a written thought – something I can go back to again and again; to reconnect with and relive. This allows God to re-create and revive the Spirit that was present when the idea was conceived as well as when it came into focus in written words. This is a slow 'darkroom' process for me. It's as if my brain is still developing film while others have moved on to the digital world. Somehow, there seems to be some correlation or connection to capturing a powerful thought on paper and stopping time with the shutter of my camera. There is a joy and peace that I feel very deeply when my words are formed into truth or a photograph captures a beautiful scene.

Sabino Canyon, Tucson, Arizona

November 8, 2010

My *false self*, the one that is still in bondage to fear (the fear of confrontation and separation) is so easily aroused. This *false self* seems to never slumber or if he does it is not a deep sleep, more like with one eye open, always willing to jump up and steer me away from a healthy response. In fact, he takes me to a *neutral* place where it feels safe and secure, a place where conflict is not invited in. It is dull in that place and all too familiar. I'm used to it. It is such a secret place that no one even knows when I am there—no one except Jan and me (sometimes). Jan knows every time I am *there*. I am so accustomed to it that I often enter in without knowing. It is such a selfish thing to do, to get comfort from *anyone* else but Jan. Yet, I have an intimacy with my *false self* that is sick. It's adultery in its most subtle attire. This relationship is killing my marriage, it's killing me.

I am amazed at how God uses me in His Kingdom in spite of this habitual *sin*. What would my life and mission be like if I was *well*? What would my marriage be like? I want to know!

Lord, forgive me, PLEASE give me the power and will to forgo my *false self*. I do not want to be afraid to deny him. Let me start today. Give me courage.

Parks, Arizona

May 4, 1991 Young Life's Trail West: silent retreat with Richard Foster

My morning prayer:

Lord forgive me for not allowing myself to admit my desperate condition. I have not cried out in desperation for your healing. Lately, when I have begun to, I have allowed those voices of rebuke to silence me. Please encourage my humility. Enable me to admit my blindness. Jesus, forgive me for not waiting by the roadside for you. Lord what do I need? **Humility**

'Lord give me humility' or 'Lord humble me'; what shall I pray? If I pray for humility maybe, over time, I'll receive it with no more effort than asking. But if I pray "Lord humble me" I'm afraid he will. I'm afraid of his method. I can't see a "nice" way to accomplish this. I'm afraid, yet I know this is what I need. Lord help me! Give me courage to pray correctly.

Maybe I will be humble immediately like the blind man who Jesus healed. That would be nice. Something inside of me is saying that I need to walk through a time, I need to pay for the journey. That it can't or won't be immediate. Maybe I should just pray and then have the faith to praise God as if it is done. I see the issue is not; what is the correct prayer? The issue is; how long will He take to achieve its end and will it be painful? I see now that it will be a lifelong process and it will require times/seasons/moments of pain. I trust all too well that Jesus will surely answer this prayer if I ask. I have never been this afraid to ask for anything in my life. "Lord not my will but yours."

The sun has shone on me just now and across the valley is a rainstorm.

The Earth needs both sun and rain. So does my life.

Lord humble me! I expect rain with the sun, there's no other way.

Water Lily

January 26, 1992

I had a discussion with Don Worcester in Prescott at the Young Life leadership weekend. He asked me how it's been going and I recounted most of the events listed below from the previous year. I made reference to my May 4th prayer at Trail West, for the Lord to 'humble me'. He paused and said, "*Wow, the Lord kicked the shit out of you!*" This is when I realized the Lord had indeed answered my prayer. (In fact he was answering it before I even asked it.)

My *humbling* year of 1991:

January: I led a seminar on *confession* at a Young Life retreat.
February 5: A low tubal transfer was unsuccessful on Jan. We decided to stop trying to *manufacture* a baby.

March: My brother James tells me he's HIV positive then asks me to perform his wedding with Ulla.

April 16: I officiated at Phil Samson's [8] funeral.

May 4: Spent two days of solitude at Trail West and decided that I needed more humility. I prayed out loud "**Lord humble me**."

May 17: I performed James and Ulla's wedding in Chris and Greg's backyard. My whole family was present together for the first time in six years, even my dad. Casey sat on his lap the entire ceremony.

June: At a committee overnight I get hammered by a number of people and come to realize that some don't trust me.

July: I finished my Masters in Theology (after 12 years).

October: We attend the wedding reception for my mom and Jess.

December 13: My dad commits suicide in Aruba.

December 19: I led a Memorial service for my dad in Tucson.

December 19: Stu and Tessa decide to move to Russia and Don Worcester decides to leave the YL staff.

December 20: A kid from the College Sunday school class we teach commits suicide. He confided in me on the 15th about his girlfriend breaking up with him.

March 31, 2013 40 year Anniversary of our horseback ride to Calgary
Jeff and I have spent several hours being interviewed by couple of reporters from the Arizona Republic/News 12 this weekend. It was fun to recount some of the stories. But, the enlightening and revealing moment came as a result of their questions that were trying to uncover our motives for going and the things we learned

about ourselves as a result. It was a good time of introspection for me.

One thing that has become very apparent to me is that we did not begin this adventure with a 'destination' mentality. We knew where we were headed; Calgary, Alberta Canada. But, the trip was never *about* Calgary. We were far too caught up with the moment by moment need to survive to ever dwell on the *finish line*. The simple fact that we are celebrating our *departure date* and not our *arrival date* is an obvious testimony to that reality. Frankly, I would need to look at my journal to determine the day we arrived in Calgary.

This is a very powerful revelation for me today to realize that the Lord impressed, infused, instilled in me a *journeyman's* heart as a result of this unique adventure. We did not arrive in Calgary with fanfare and a big celebration. Our last day of this amazing journey did not *feel* any different than the first. This is an obvious sign that the 'journey' was clearly the point and not the 'destination'. Before today, I have never made this connection. I don't know why the Lord has kept this "secret" from me until now – especially after all of the talks I have given over the years on *journey vs. destination*. But, it confirms and validates the message. I have embraced and believed this 'truth', and passionately spoken about it these past few years without realizing that the Lord had secretly (profoundly) cemented it in my heart 40 years ago.

My prayer *theme* this year has been to ask the Lord to help me *stay in the present moment*, to not be anxious about tomorrow. It could now be restated; "Lord, renew in me the heart I had at 19, that was able to faithfully live one day at a time walking in the direction you had charted out before me unconcerned with the finish line".

60 miles from Calgary, Alberta, Canada, 1973

November 21, 2003

I'm sitting on the west end of the island of Aruba. I believe I am very near the place where my dad took his life almost 13 years ago. It is desolate and beautiful. The earth is hard without much vegetation. The water is multicolored, teeming with life. What was going through my dad's mind as he drove to this place, to his death? His journey was a dichotomy: always pursuing the most beautiful place or vista but coming to it so lost and arid. As if the beauty and mystery of that location would heal him or bring him peace. But, places (even the best ones that God took extra care to fashion) cannot heal us for long. It is just momentary: a peaceful hour, and a time of introspection and then back to the pain of being you. My dad sought out beautiful places hoping to find *the* place. But, creation cannot heal us, only the Creator is able to do that. Why did my dad fear God so much? What happened to him along his way? As he looked over this awesome scene – the Lord must have been speaking to him. Why didn't he, why couldn't he hear? In my dad's last hour did Jesus come to him? Did he finally succumbed to the Lord or in the end just succumb? It is finished. How beautifully empty. Lord, may I never miss your '*still small voice*' as I ponder and view your creation. May I always know my Fathers' love in the moments when I am alone and looking upon a vast horizon. I am so blessed, my life is so rich. I am loved by God.

Puerto Penasco, Mexico

*November 2012 The old "bait and switc*h"

*Toroweap: I went for sunsets—and I got lightning!

*Beyond Malibu: I went on a hike with some kids—I met a woman who would become my wife.

*We expected to raise a child—and the child raised us.

*I often thought I was taking Jesus to a place—only to find He was already there.

I don't know why I still allow myself to think I have anything figured out. I don't think I am ready to throw away my calendar (my plans) and just wait each morning for my daily instructions from God. Please don't misunderstand me, I absolutely believe that God could handle that approach—in fact he may prefer it. (It worked well for Jesus.) My genetic circuit board is wired for control. But, I believe after hundreds of obvious holy intrusions I find myself beginning to loosen my grip a little. Less expectations and agenda, more wonder.

Midnight Monsoon, Toroweap, North Rim of the Grand Canyon

April 29, 2015

I am reading Thomas Merton today. He said, "The desert was the region in which the chosen people had wandered for 30 years, cared for by God alone. They could have reached the Promised Land in a few months if they travelled directly to it. God's plan was that they should learn to love him in the wilderness and that they should always look back upon the time in the desert as the idyllic time of their life with Him alone."

This causes me to think about how so many people, including me, *look back* at their early days of adulthood or marriage and recall; 'we didn't have much', 'that first apartment was so small', 'we only had one car'. It was such a simple season which required a lot of faith to make ends meet. I look back at those days and often long for that 'simple season' where we were conscientious of what we *needed* vs. what we *wanted.* When 'generic' was always the first choice of products and splurging meant we'd buy the *big* pot for the patio and fill it with a *whole flat* of flowers. Faith and intimacy was nurtured in this 'wilderness'. Every decision was discussed and evaluated. When you don't have many resources you are forced to weigh the consequences of your choices. Our *normal* dialogue nurtured intimacy and molded our hearts. *Wilderness* makes one needy and forces you to reach out, cry out to God and find counsel in another. I wonder if when I look back nostalgically on those early days if I am not longing for the simple intimacy that we enjoyed as we navigated that season?

New Year's Day snow, Scottsdale, Arizona

December 16, 2014

The other day there was a weird cloudburst of rain over our house. It was late afternoon on a sunny day and, out of nowhere, I hear thunder and this storm shows up and begins to rain on our neighborhood. But there was sun shining to the north, south and west. The little storm seemed to settle over central Phoenix and Scottsdale. We were going to Scottsdale so I took my camera in anticipation of a rainbow. Almost immediately a very bright arc appeared below the rain clouds as bright sunlight from the west fueled the spectrum in front of us. As we drove past the North Mountain Preserve the rainbow kept its distance out in front of us as its ends washed over hills and valleys painting everything it touched in true light. We stopped so I could take some photographs. I noticed when the rainbow settled on the top of the hill there was a hiker in a red shirt. He/she was being drenched in the full spectrum of rainbow light. It was such a glorious sight. I thought, how cool would it be to stand on a hill in the rain and be illuminated by pure rainbow light. But, that's impossible because we can't catch up with rainbow light, it is always on the move away from where we are. So, the person in my photograph whose red shirt is being exposed by the rainbow isn't really in the light, at least from his perspective. He's looking out at the road I am on being blessed by the sight of a rainbow that's just beyond his reach. We both see each other in the light of the rainbow but neither one of us is actually in it. Or, we are both simultaneously in it? Or—? (I'm sure a very bright mathematician or astrophysicist might attempt to explain that weird formula that proves the closer you get to the speed of light the slower time goes by and therefore a rainbow can be at two places at the same time and in neither place at all—)

I want to stand on the top of a hill and be washed in rainbow light. I want to raise my arms to the heavens and absorb the full spectrum of God's creation we call light. It seems like one would never recover from that illumination – that at some level it would alter you in some holy, secret, cosmic way. On the other hand, since we know that light contains all of the colors of the rainbow, I would have to say I am being washed every minute of every day in glorious rainbow light. I think the secret, the challenge, is to stop on a regular basis and raise my arms to the Lord and be washed in His pure holy light knowing, that from his perspective, I am indeed being illuminated by his magnificent infinitely colorful light! No need to wait for a storm because He is the Light of the world, all of the time!

North Mountain Preserve, Phoenix, Arizona

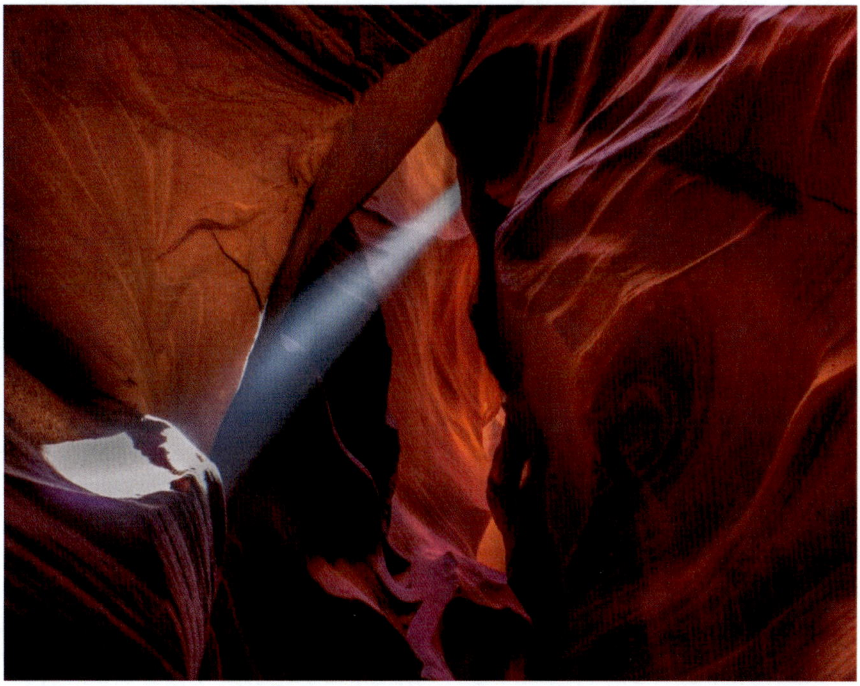

Upper Antelope Canyon, Arizona

October 15, 2013 *The Lord spoke to me in Upper Antelope Canyon*

I've always wanted to go and experience this amazing place – a slot canyon winding through multicolored sandstone on the Navajo Reservation where beams of sunlight (depending on the time of year and time of day) will shine into this stone sanctuary and create surreal almost heavenly sights.

We reserved a guide, Ray, and he led us into the canyon with three very specific strong instructions: *stay with me, move when I say move, stop when I tell you. I'll get you the shots you want but you have to stick with me. Many times we will only have a minute to catch an image – it's all timing...*

The canyon is popular and always busy but, this weekend was particularly full because of the government shutdown had closed all the National Parks including Lake Powell. There were tourists from all over the world trying to find anything possible to fill their time. Consequently, the slot was full of people, wall to wall.

Since we were a *photographic group*, Ray could muscle his and our way through the crowd and allow us to set up our tripods in a hurry, snap a shot, and then move on. We had two hours of "let's go, let's go, time to move, please folks please wait while my group takes their photos." There we were in this amazing castle of light and color and motion and the whole experience felt like a 'track meet' interspersed with pauses to record heavenly sunbeams and sand falls.

Near the end my camera told me that my 'card was full' *—no more memory*! I couldn't take any more pictures! My other memory card was in the car. Bummer, there I stood unable to *do* anything. I was forced, by the Lord, to actually just stand and look around and for the first time see, feel and experience this sanctuary.

In the blink of an eye my Heavenly Father spoke gentle words, he shown a beam of his Holy light onto the 'floor' of *my* canyon and exposed my current condition. I have been in too big of a hurry lately. He has been leading me through some amazing places alongside some amazing people and I have missed most of the *scenery* because I was in too much of a hurry to get to the next place so I could capture the perfect 'photo'.

The past two hours were a short 'documentary' on the last few months of my life. Not until my 'card was full', when I had no more memory and couldn't *do* the thing I came there to do, was I able to hear my loving Father's graceful reprimand;

Stop and be still! *You are hurrying through your journey, passing by my people and my beauty. The journey I have scheduled you for is world-class. But, it will only produce a few nice snapshots if we don't travel it together. I filled your memory card with images. Stop and let me fill your soul with myself.*

How long will it take for me to get *this picture*? I am grateful for God's patience.

Upper Antelope Canyon, Arizona

9

RELIGION! HEAVEN HELP US

July 27, 2009

I read in the newspaper about a man in Pakistan who heads up 'The Movement for the Enforcement of Islamic Law'. I'm trying to get my mind around the idea of enforcing religious laws. I can see clerics or theologians teaching about their laws or holy books but to enforce the truths they believe in seems weird. Forcing people to believe or follow the rules of his/her religion seems counter to one choosing the teachings as a way of life. When does choice become an option?

Is the church (religious people) afraid of choice? Is it that we don't trust people to make the right choice so we make them believe? Isn't the problem more about our trust in God? Maybe we don't believe that God's truth can stand the test of our children or of the congregation. So, we learn the subtle art of not allowing for searching and testing. The end result in the Christian church is often a congregation of shallow passionless people. For those under Islam and other faith traditions the result is often the same: religious rules abiding people going through the motions without passion and consequently joy.

Bethlehem, Palestine

July 29, 2010

I read an editorial titled: *What happened to free time?*

The author hits on a question/message I have been pondering for a long time. I remember back in the 70s when it was predicted that the world was going to change because of the personal computer. We would become so much more efficient that our free time would increase and leisure would become more available. Indeed, today we are painfully efficient but what a swing and a miss by the culture kings – the sociologists who thought humans valued leisure or free time.

If free time is indeed a "value" in the human psyche it has been handcuffed, chained up and treated like an ailment. Today, if you have too much free time the general population will look upon you with disdain – or at least assume that you are lazy, unmotivated. Certainly the church doesn't honor or promote leisure. We teach ourselves how to manage our time in a holy way: spend 'x' amount of hours with the kids and have a regular date with our spouses, and don't forget your obligations at the church.

Rest seems like a good idea as long as we stay productive and keep up. This sense of competition, of measuring up, is out of control. The speed at which we can move through information is making us crazy. It seems unnatural. The ethic of Sabbath, rest and non-productive time is way down the list of important values in our modern Western society. Why don't we congratulate the person who says, "Thanks to my PC and the ability it provides me to speed up my work, I now have 4 day weekends."? Instead, with well-developed subtlety, we condemn him/her for not using the 'gift' of time to produce more. I know I am a product of this mentality. I am guilty of judging others as well as not giving myself permission to 'waste time' because of how I fear I will be perceived. What would Jesus say about this?

I need to ask him – while I'm wasting some time...

July 31, 2007 Literally uneducated

A thought about taking our Holy book(s) literally:

As I was reading in Luke this morning I came to an interesting, if not disturbing, passage:

Then Jesus asked them, "When I sent you without purse, bag or sandals, did you lack anything?" "Nothing," they answered.

He said to them, "But now if you have a purse, take it, and also a bag; and if you don't have a sword, sell your cloak and buy one. It is written: 'And he was numbered with the transgressors'; and I tell you that this must be fulfilled in me. Yes, what is written about me is reaching its fulfillment." The disciples said, "See, Lord, here are two swords." "That is enough," he replied. Luke 22:35-38

Here we find Jesus warning his friends of the turmoil that is about to overwhelm them. He tells them if they do not have a sword to buy one. After they say, "here are two swords", he says, "That's enough."

Now I am thinking to myself that this is a very heavy statement. Does Jesus the man of peace who is the embodiment of grace want his friends to take up arms and fight for him? I went to some of my literary tools and discovered that in the original Greek (according to a commentator) the meaning is clear:

It does not mean; 'Two will be sufficient' but rather, 'enough of this kind of talk!' This is a way of dismissing a subject in which the disciples were hopelessly astray.

This may be clear to very learned scholars/commentators but not so for the casual reader. For someone who has not been educated or discipled, or has no evangelical church background, this passage (not to mention many others) is quite confusing and disturbing. Taken literally, it can cause some real turmoil. But, over the ages, many wise men and women scholars have helped illuminate the texts and consequently enabled us be comfortable with our book.

Our dear Muslim friends also have a book (The Quran). For them it is the most Holy of all books. The world is looking hard at their book these days because many evil men are taking some of its words literally and using them to give themselves permission to kill innocent people; Muslim, Jew, Christian and atheist alike. Many people in the West believe these words are literal as well and have determined that all who follow that book are evil. The problem here is that in Islam the Quran is so holy that it is unlawful/forbidden to analyze it, to explain it or critically comment on it. So the hard and troubling texts in it are forced to stand on their own two literal feet. I am so thankful for the history of biblical exegetes (even the wacky ones) who have gone before me to help make my book understandable. What a gift to me and my journey with God and my understanding of Jesus.

Phoenix, Arizona

August 26, 2013

Tom Melton spoke about light yesterday from Matthew 5:14-16. He used a lighthouse as a metaphor. Jesus said you are light, not become light. I started thinking about a lighthouse. Its job is to warn about the land and also to invite or welcomed the weary mariner. It seems that many in the evangelical tradition have been taught to warn more than to welcome. We have learned to use our light to scrutinize, expose and judge more than show the way, to welcome the weary traveler ashore.

10

GRACE IN TWO LETTERS

June 12, 2011

Dan,

I so appreciated our time yesterday. I'm humbled by your honesty, your trust in me. God is moving (as always) and it seems he has been getting your attention lately even before your daughter's issues surfaced. The fear and concern around her tumors illuminated His presence, His voice. I want to comment on two things you shared with me yesterday. One has to do with a 'hole' in your life and the other with the burden of guilt you are carrying.

I'm convinced the Lord is speaking to us all of the time and sometimes we're actually listening. In a quiet moment of clarity in a hotel room in Los Angeles you looked out of a window and saw yourself. You heard God's voice when you gazed at that building with a hole through it and thought, "I have a hole in my life". Your building is strong, great architecture, but something is missing. So, you do what most of us would do; default to making a list of holy things to fill it with to appease God: good work, service projects, go to church more... It's an age-old response. In the 6th chapter of the Gospel of John we see what Jesus had to say to those who wanted to know what to "do": Then they asked him, "What must we do to do the works God requires?" Jesus answered, "The work of God is this: to believe in the one he has sent."

Dan, your natural instinct is to fix, to work, to make it happen. Naturally, you want to fill the hole with something. But God does not want you to do any work, to offer a sacrifice or to do anything. What God wants from you is for you to believe in him and to trust him. That's how to fill the hole.

I'm so sorry that you're burdened by some past guilt. Many people are. You said that you really loved the *Prodigal God*[9], that it was the best book our group has ever read. Why? Because it gave a very clear picture of grace—grace that you and I crave. Grace is an encouraging concept, an amazing theological idea. It's hard to imagine that our Heavenly Father loves us in spite of our secret sins and blatant disrespect for him. But, the Gospels couldn't be any clearer; He loves us regardless of our story and our choices. His amazing love is such an indescribable and unbelievable concept. Unfortunately, it often remains just a 'concept' for many of us because we

can't, won't, don't allow His undeserved grace to be true for us. We believe it as a theological truth. We are amazed by its implications and transformative power in the lives of some people we know. But, all too often that is where grace ends up: as an idea in a book or a transformative miracle in someone else's life. Dan, to make grace a reality in your own life requires humility, courage and trust. Like the "prodigal' you need to acknowledge your sin intellectually and emotionally. You need to identify it (maybe write it down like the son in the Luke 15 account) and then go home to your Father where He is waiting with open arms to welcome you. This courageous humiliating time of confession will free your heart and soul and be a watershed moment.
Go home my friend and get loved... Your brother in Christ

October 14, 2002

Jay,

Sorry for the delay in my response. I have been on the road a bit.

Sounds like the whole "selfish" thing has got you on the run. Welcome to the human race. I wish I could wave a magic wand and free you up from that burden/prison but I cannot. I live with the same tension. It's chronic and can be fatal. Here are some thoughts your note has provoked in me:

I believe we need to be careful to not make a habit out of listing and then dwelling on our failures (sins). That in itself can be selfish (narcissistic) and depress the hell out of us. This does not mean we deny our sin. It is real and can kill us. I just think it is healthier to view ourselves as loved and forgiven; saved by grace (Eph.2:8-9). Then by faith (true humility) we choose to stay in touch with Jesus moment by moment. Wallow in his extravagant love asking for his forgiveness in the moment. I believe if we commune with him, walk with him, serve him in the present, our future (which is his) and our sin (which is past, done, unchangeable) will not burden us. It is His burden. His yoke upon us is not heavy (Matt. 11:29-30). You and I burden ourselves with what it will be like after work, in my next career, next season—I believe we are called to live in the now, present, ever looking for the Lord in our current "being". Look at the Gospel of John and count how many times Jesus said 'I only do what the Father tells me to do.' He rarely looked back or forward. When he did (i.e. in the garden) he quickly said, "Not my will but yours be done." He has so much to teach us in the present. We cannot learn from what has not come to pass. The Prodigal learned a lot from feeding pigs. He never dreamt of partying, just surviving.

Jay, wallow in your sin. Reach out to the only one who cares; the one who loved you before you were born, who thinks you're worth dying for. He is the one who knew you as a child in a broken family. He knows all about why you lust and whom and what you lust over. He sees you as his favorite son, not a hopeless sinner. Walk with your Father today. He is unconcerned with who you become because he is so excited about who you are. Allow Him to embrace you today. He waits for your invitation. The thief on the cross asked Jesus, "Remember me when you come into your Kingdom." Jesus answered him, "I tell you the truth, **today** you will be with me in Paradise." Jay, Jesus exists for us today, not someday when we are good enough. He wants to walk with thieves today. Let him hold your hand like the Abba that he is today. Don't wait. The future becomes tomorrow so quickly when we live in the now of the Savior.

Keep seeking the Father. May you know his embrace today my friend.

11

FREE ADVICE

October 24, 2007

My thoughts have been prompted by the fires in California. Drought, wind and flame, an explosive combination in the physical world; Might there be an application in our spiritual life?

What happens when we can have a dry season internally – in our soul, where the spirit has not rained (or reigned) for a while? The warning signs go up along the road/path: **Fire Danger High**, no smoking or off-roading allowed. The strong wind blows. It's predictable, even expected. Under different circumstances it can be pleasant. It causes wind chimes to sing and cools our brow. But when it blows over parched land the warning signs go up. As the old song goes: "It only takes a spark to get a fire going." But the fire that burns under these circumstances does not invite us to "warm up to its glowing." That kind of fire is a wildfire that can get out of control and consume acres of our soul where fuel has been accumulating. We must be diligent in our journey with the Lord. We cannot get lazy and forget to water/saturate ourselves with his cool water—His presence.

Currently I can see smoke and flames of wild fires all around me – in the lives of many friends who have neglected going to the Spring. Now, sparks have been discarded near their roads and have fueled devastating fires consuming every good thing in their path.

Lord, wash me in your holy water, keep the hills around my soul green and alive. Cause me to enjoy and not fear the sound of strong winds. Put your strong heel to the discarded matches and cigarette butts that get discarded along my road. Rain on my soul. Reign in my heart.

Angels Window, North Rim Grand Canyon, Arizona

April 21, 2012 Marriage 'rest stop'

So you stop along the way, along your marriage, and you wonder; is this it? Is she/he the right one? Everyone has made that stop – pulled over to the side of the marriage road and wondered. If not yet, you will.

When you do, you may cry out loud, really loud! Or, have such a faint thought it doesn't even rate a whisper. But, for some reason, you will give your mind permission to wander and wonder. The place where you chose to pull over can greatly influence the path your thoughts walk down.

If you pull into an official rest area with neon lights, restrooms, vending machines and picnic tables, chances are you will be distracted. A man-made oasis is unnatural and inhibits our senses because it limits our choices. Its 'limitation' is a consequence of its choices: Coke or Pepsi? Men's or Women's? When you feel the need to pull over on the marriage road to ponder and think, make sure it's a place where there is gravel under your tires, void of others, and has a clear view of the horizon. This is the place you want to be when you ask if he/she is the 'one'. Here is where your only distractions will be natural, where the only things you see and the sounds you hear will be created by God. This is the place where Truth will have a chance to pour into your introspective moment, your heart. This is where you will be comforted not condemned for making the stop. Make sure you stay long enough to hear His still small voice.

His loving words to me about Janny during my stops have been more like an aroma – sweet incense that have enabled my heart to conjure up holy memories...

*an unlikely encounter in the Canadian Rockies
*a peace about our future that clearly surpassed all understanding
*adventures that few women would ever want to take
*a real lover who told me the truth
*holy grace that tells me to keep going
*a miracle child that couldn't have been formed inside any other womb
*a strong faith that won't give up on Jesus or me

These are a few of the strong words that I have heard during my rare pullovers.

Don't be afraid or ashamed to stop. But, be warned! Choose your rest stop with great care. Avoid the shiny: the secure paved place with fabricated options and illusions. Choose a natural and quiet place where you have a view. Where earth touches your tires and Jesus is in the breeze. This is where your First love will remind you of moments with your true love that He has ordained, blessed and given you as a precious gift.

EPILOGUE

October 10, 2007 Remembering July 2006: trying to keep things in focus[10]

So bombs fly and people run. They're just people – people without names in a part of the world that most cannot locate on a map. Fear and death and sadness in an obscure place involving people with names we can't pronounce, like we even care to try. Families, homes and towns destroyed—dreams and hopes fade away. But, distance, time and ignorance keep us immune from caring. It's just another CNN story sandwiched between advertisements. Another day, another bomb. Another day, another sound-bite. Another day, another tragic story in a distant land involving people you don't care about.

You don't care unless they are people you call friends. People with a name you've learned to pronounce from a town whose streets you have walked and whose table where you have shared a meal. It's just another story until your friends flee their home and leave their table and run from a place you know, a place you can picture, holding tightly the hands of children you care about. It's a deeper pain and a truer caring when the people have a name you have learned to pronounce and a story you know in part.

And then the awful reality of a deeper cause for caring 'blows up', is exposed when your own flesh and blood is a player in the story. One whose name you can pronounce because you gave it to her. One whose home and table and family are too real because they are yours. That moment of awful reality when CNN becomes your comfort, your lifeline to your daughter. This explosive day could not have been predicted, expected or much less desired. That day when nowhere became somewhere, when nobody became somebody, when a friend became family. This is the day God got bigger, my faith got tested and my world was expanded. I would have voted (if a vote was taken) to skip this test. Thankfully, no one asked for my permission or input. Consequently, fear and anxiety prevailed—and blessings flowed. My map was expanded, my family grew larger, my love was exposed and my God got bigger.

Dear sovereign Lord,

Thank you for your grace and love. Thank you for a place called Lebanon and all other places. Thank you for Fouad, Gina, Chadi, Mona and every person on earth with a name I cannot pronounce. Thank you for the Orient Queen and all other ships that rescue people. Thank you for Janny and Cassia – the best names in all of the world.

May your hand prevail on earth so that families in every town on your map can live without fear and in peace. Amen

Southern Beirut, Lebanon Fall 2006

November 6, 2013

Thoughts about children are really ringing in my ears these days. I am sure Charlotte has awakened my receptors. G.K. Chesterton writes, "It is possible that God says every morning, "do it again" to the Sun; and every evening, "do it again" to the Moon. It may not be automatic necessity that makes all daisies alike; it may be that God makes every daisy separately but has never gotten tired of making them. It may be He has eternal appetite for infancy; for we have sinned and grown old, and our Father is younger than we."[11]

I don't want to lose the childlike joy that Charlotte has. I should say, I want to regain it. Graham Greene writes, "There is always one moment in childhood when the door opens and lets in the future."

Charlotte's dinosaurs running wild at the Grand Canyon

Lord, surround Charlotte with joy – hold the future back as long as possible.

WORKS CITED

1-Duncan, David James. *My Story As Told By Water*. (Sierra Club Books in conjunction with Crown Publishers, New Yor, New York, 2001(, 50-51.
2-*Lost Canyon* is a *Young Life* camping facility in Williams, Arizona
3-Eldredge, John. *Wild at Heart*. (Thomas Nelson, Inc., Nashville, Tennessee, 2001), 31.
4-McDonald, George. *Phantastes*. 1858.
5-Bell, Rob. *Velvet Elvis*. (Zondervan e-book, Grand Rapids, Michigan. 2005). 131.
6-*Canaan in the Desert* is a retreat center founded the *Evangelical Sisters of Mary* in Scottsdale, Arizona
7-Gire, Ken. *Windows of the Soul*, (Zondervan, Grand Rapids, Michigan 1996), 187.
8-Phil Samson was my best friend's dad. He stepped into my life when my father stepped out. Phil showed me what a good father was like.
9-Keller, Timothy. *Prodigal God*. (Penguin Group, New York New York. 2008).
10-In the summer of 2006 our daughter Cassia had gone to Beruit, Lebanon to serve with friends who were caring for people in a Palestinian refugee camp. Three days after she arrived, Israel started bombing the city.
11-Chesterton, G.K. Orthodoxy. (Image Books, A division of Doubleday & Company, Inc. Garden City, New York. 1959), 60.
.

THE AUTHOR

W.J. (Bill) Hamilton is an Arizona native. Jan, his wife of 38 years, Cassia, his daughter, Brad, his son-in-law and Charlotte and William, his grandchildren, are the highlights of his life!

The first half of his vocational life he served in the mission of *Young Life*. In the second half, by the grace of many friends, he has been set free to be available to walk alongside and encourage people in their spiritual journey/pursuits.

Bill's avocation is photography. He does not consider himself an artist but a 'delivery boy'. In his view, all of creation is a work of art that has been *painted* by God. Occasionally, he has the privilege of being in the right place at the right time to frame one of His masterpieces.

You can contact Bill Hamilton at : billhamiltonaz@gmail.com

Made in United States
Orlando, FL
04 November 2026

90202063R00069